Common Core Math Literacy

Assessment Prep for Common Core Mathematics

Tips and Practice for the Math Standards

Grade 6

AUTHOR: KARISE MACE and CHRISTINE HENDERSON
EDITORS: MARY DIETERICH and SARAH M. ANDERSON
PROOFREADER: MARGARET BROWN

COPYRIGHT © 2015 Mark Twain Media, Inc.

ISBN 978-1-62223-529-2

Printing No. CD-404232

Mark Twain Media, Inc., Publishers
Distributed by Carson-Dellosa Publishing LLC

Visit us at www.carsondellosa.com

Table of Contents

Introduction to the Teacher

The time has come to raise the rigor in our children's mathematical education. The Common Core State Standards were developed to help guide educators and parents on how to do this by outlining what students are expected to learn at each grade level. The bar has been set high, but our students are up to the challenge.

More than 40 states have adopted the Common Core State Standards, and the school districts in those states are aligning their curriculums and state assessments to those standards. This workbook is designed to help you prepare your students for assessments based on the Common Core State Standards. It contains both multiple-choice and open-ended assessment questions that are similar to the types of questions students will encounter on their state assessments. We have also included test-taking tips and strategies that will help students perform well on these types of assessments.

Additionally, this book contains diagnostic information for the multiple-choice questions. This information will help you understand why your students selected particular incorrect answers. We believe that you will be able to use this information to identify the gaps in student knowledge, which will inform your future instruction.

We hope that this book will be a valuable resource for you in preparing your students for assessments that are aligned with the Common Core State Standards!

—Karise Mace and Christine Henderson

Test-Taking Strategies for Math Tests

Test anxiety affects many students. Here are some strategies you can teach your students to help alleviate the anxiety and help them become more relaxed test takers. We have also included a sample problem in which we highlight how these strategies might be used.

Multiple-Choice Tests

Tip #1:   Read the problem thoroughly and determine the goal.

Anxious test takers have a tendency to read through problems quickly and then immediately scan the answer choices for what might be the correct answer. Encourage your students to be patient as they read through each problem so that they can determine what the problem is asking them to do. They may even wish to circle information that they think is important and underline the question.

Tip #2:   Estimate the answer.

Students often "number surf." That is, they "grab" the numbers they see in the problem and start operating on them in an attempt to get one of the answer choices. Encourage your students to use estimation to determine the reasonableness of an answer.

Tip #3:   Use your estimate to quickly eliminate one or two of the choices.

Once students have calculated an estimate, they can almost always use it to eliminate one or two unreasonable choices. Encourage them to cross these out with their pencils.

Tip #4:   Solve the problem by working forward or backward.

Some problems can be solved just as efficiently by working forward or backward. If students are unsure about how to use the information in the problem to get one of the answers, encourage them to start with one of the answers and work backward to see if they get the information in the problem.

Example: You have 12 yards of ribbon. It takes $\frac{2}{3}$ of a yard of ribbon to wrap a package. How many packages can you wrap?

    A. 24 packages
    B. 18 packages
    C. 16 packages
    D. 12 packages

Estimate: I know that $\frac{2}{3}$ is more than $\frac{1}{2}$ but less than 1. So, the number of packages must be between **12** and **24.**

Eliminate: Because the number of packages must be between 12 and 24, I can eliminate choices **A** and **D.**

Working forward:

12 yards $\div$ $\frac{2}{3}$ yard/package =
    $\frac{12}{1}$ x $\frac{3}{2}$ = 18 packages

The correct answer is choice **B.**

Test-Taking Strategies for Math Tests

Open-Ended Response Tests

The tips for solving open-ended response problems are similar to those for solving multiple-choice problems. However, because open-ended response questions are also used to assess the problem-solving process, students must learn how to communicate their process. These tips will help them learn to do that.

Tip #1:   Read the problem thoroughly and determine the goal.

Open-ended response problems are often multi-step. It is important to encourage your students to read these problems patiently and thoroughly so that they do not forget to complete the problem. It may be helpful for them to circle important information and underline the question.

Example: Maggie has 110 feet of fencing and would like to use it to enclose a rectangular area that is 32 feet by 25 feet. Does she have enough fencing to do this? Explain your reasoning.

Tip #2:   Make a list of what you know and what you need to figure out.

Making lists can help students keep their information organized. Encourage them to make two lists—one of the things they know and another of the things they need to figure out.

Things I know:
1. Maggie has 110 feet of fencing.
2. The area to be enclosed is a rectangle.
3. The length of the rectangle is 32 feet, and the width is 25 feet.

Things I need to figure out:
1. What is the perimeter of the area to be enclosed?
2. Whether or not Maggie has enough fencing to enclose the area

Tip #3:   Devise a plan for solving the problem.

While students do not always need to write out their problem-solving plan, it is important for them to form one. Many open-ended response problems ask students to explain their problem-solving process. Encourage students to write down their plan as part of this explanation.

Plan: I am going to calculate the perimeter of the rectangular area and compare it to the amount of fencing Maggie has.

Tip #4:   Carry out your plan.

As students begin to carry out their plan, encourage them to show their work!

Tip #5:   Check your work.

Students like to skip this step, but it is one of the most important ones in the problem-solving process. Encourage your students to take time to check their work and to make sure that they actually solved the problem they were asked to solve.

Carry out the plan:

$P = 2l + 2w$

$= 2(32) + 2(25)$

$= 114$

The perimeter of the rectangular area is 114 feet. Maggie does *not* have enough fencing to enclose it because she only has 110 feet of fencing.

Geometry

Problem Correlation to CCSS Grade 6 Geometry Standards

MC Problem #	6.G.A.1	6.G.A.2	6.G.A.3	6.G.A.4
1	•			
2			•	
3		•		
4				•
5				•
6		•		
7			•	
8	•			
9	•			
10				•
11		•		
12			•	
13		•		
14	•			
15	•			
16			•	
17			•	
18		•		
19				•
20				•
Open-Ended Problem #	6.G.A.1	6.G.A.2	6.G.A.3	6.G.A.4
1	•			
2		•		
3			•	
4		•		•

Name: _____ Date: _____

Geometry: Multiple-Choice Assessment Prep

Directions: Circle the choice that best answers the question.

---

1. What is the area of the shaded triangle?

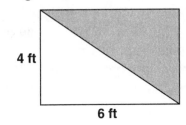

A. 24 ft²

B. 24 ft

C. 20 ft

D. 12 ft²

---

3. What is the volume of the cube?

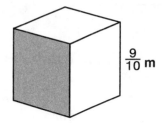

$\frac{9}{10}$ m

A. $\frac{27}{10}$ m³

B. $\frac{27}{30}$ m³

C. $\frac{27}{1000}$ m³

D. $\frac{729}{1000}$ m³

---

2. If the given points below are plotted on a coordinate plane, which of the following best describes the resulting polygon *ABCD*?

$A(-3, -6)$, $B(-3, 1)$, $C(7, 1)$, and $D(7, -6)$

A. Parallelogram

B. Square

C. Rectangle

D. Trapezoid

---

4. Which answer best describes the three-dimensional figure that can be formed by the given net?

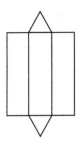

A. Pyramid

B. Triangular prism

C. Prism

D. Cone

Name: _____ Date: _____

Geometry: Multiple-Choice Assessment Prep

Directions: Circle the choice that best answers the question.

---

5. What is the surface area of the three-dimensional figure formed by the given net?

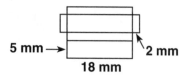

5 mm →
2 mm
18 mm

A. 136 mm²

B. 180 mm²

C. 272 mm²

D. 380 mm²

---

7. On a map of a town square, the library is located at (4, 5), the post office is located at (4, 16), and the grocery store is located at (15, 5). The hardware store is 11 units from the post office and 11 units from the grocery store. What are the coordinates of the hardware store?

A. (26, 5)

B. (4, 27)

C. (15, 16)

D. (16, 15)

---

6. The right rectangular prism shown below has a volume of $\frac{5}{16}$ cubic yard. What is the area of the base of the prism?

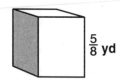

$\frac{5}{8}$ yd

A. $\frac{25}{128}$ yd²

B. $\frac{25}{128}$ yd³

C. $\frac{1}{2}$ yd²

D. $\frac{1}{2}$ yd³

---

8. What is the area of the isosceles trapezoid shown below?

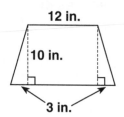

12 in.

10 in.

3 in.

A. 180 in.²

B. 150 in.²

C. 120 in.²

D. 28 in.²

Name: _____ Date: _____

Geometry: Multiple-Choice Assessment Prep

Directions: Circle the choice that best answers the question.

9. How much material is needed to make the kite that is shown?

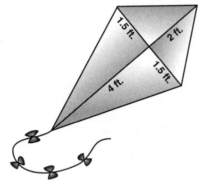

A. 8 ft²

B. 9 ft²

C. 18 ft²

D. 48 ft²

11. What is the volume of the right rectangular prism?

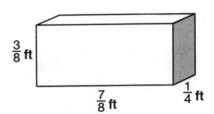

A. $\frac{21}{256}$ ft³

B. $\frac{21}{256}$ ft²

C. $\frac{3}{2}$ ft³

D. $\frac{3}{2}$ ft²

10. How much paper would be needed to cover the three-dimensional figure formed by the given net?

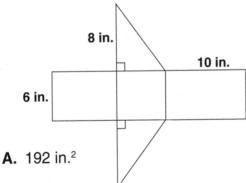

A. 192 in.²

B. 192 in.³

C. 480 in.²

D. 480 in.³

12. What is the length of side *PQ*?

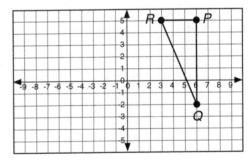

A. 3 units

B. 5 units

C. 7 units

D. 9 units

Geometry: Multiple-Choice Assessment Prep

Directions: Circle the choice that best answers the question.

---

13. The aquarium below will hold $16\frac{1}{4}$ cubic feet of water. How deep is the aquarium?

5 ft   **2$\frac{1}{6}$ ft**

A. $176\frac{1}{24}$ ft

B. $7\frac{1}{2}$ ft

C. $3\frac{1}{4}$ ft

D. $1\frac{1}{2}$ ft

---

15. What is the area of the shaded triangle?

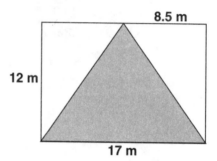

8.5 m

12 m

17 m

A. 204 m²

B. 102 m²

C. 58 m²

D. 51 m²

---

14. What is the best estimate for the area of the hexagon?

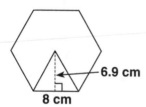

6.9 cm

8 cm

A. 48 cm²

B. 56 cm²

C. 168 cm²

D. 336 cm²

---

16. What is the length of side *ZY*?

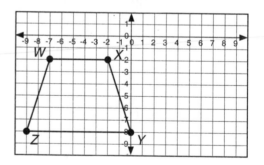

A. 5 units

B. 6 units

C. 8 units

D. 9 units

Name: _____ Date: _____

Geometry: Multiple-Choice Assessment Prep

Directions: Circle the choice that best answers the question.

---

17. Where should point *H* be placed so that *EFGH* is a parallelogram and *GH* has a length of 8 units?

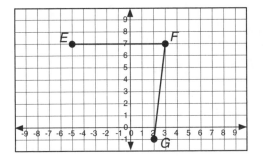

   A. (−6, −1)

   B. (−1, −6)

   C. (8, −1)

   D. (2, −9)

---

19. Which of the following could *not* be the net of a rectangular prism?

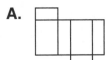

   A.

   B.

   C.

   D.

---

18. What is the volume of the right rectangular prism with a base of $8\frac{2}{5}$ square centimeters and a height of $1\frac{5}{6}$ centimeters?

   A. $15\frac{2}{5}$ cm³

   B. 14 cm³

   C. $10\frac{7}{30}$ cm³

   D. $6\frac{17}{30}$ cm³

---

20. Will 400 square inches of wrapping paper be enough to wrap the package shown below?

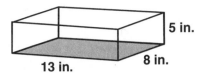

   A. Yes, because the volume of the pris⌐ is less than 400 in.³.

   B. Yes, because the surface ⌐ prism is less than 400 i⌐

   C. No, because the v⌐ is greater than ⌐

   D. No, because⌐ prism is grea⌐

Name: _____ Date: _____

Geometry: Open-Ended Response Assessment Prep

Directions: Answer the question completely. Show your work and explain your reasoning.

Problem 1: Addison is making a flag of the Czech Republic for a class project. A sketch of the flag is shown. How much red, white, and blue material will Addison need?

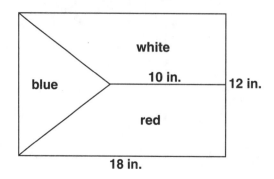

Show your work.

Explain your reasoning.

Name: _____ Date: _____

Geometry: Open-Ended Response Assessment Prep

Directions: Answer the question completely. Show your work and explain your reasoning.

Problem 2: A store is going out of business and is running a special close-out deal on their office supplies. Customers can buy Box A or Box B for the prices shown. Then they can fill the box with merchandise. Which box is a better deal? Explain how you determined your answer.

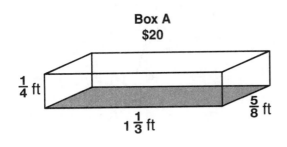

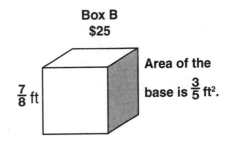

Show your work.

Explain your reasoning.

Name: _____ Date: _____

Geometry: Open-Ended Response Assessment Prep

Directions: Answer the question completely. Show your work and explain your reasoning.

Problem 3: Use the graph below to answer the questions.

1. What are the coordinates for four different points that are 6 units from point *C*?

2. Which, if any, of these points could be connected to points *A* and *C* such that figure *ABCD* is a parallelogram?

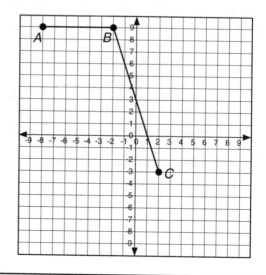

Show your work.	Explain your reasoning.

Name: _____ Date: _____

Geometry: Open-Ended Response Assessment Prep

Directions: Answer the question completely. Show your work and explain your reasoning.

Problem 4: Eleesha is making small boxes for party favors. She wants to use as little cardboard as possible. However, she would like the boxes to hold as much as possible. The nets for her two design ideas are shown below. Which one should she choose?

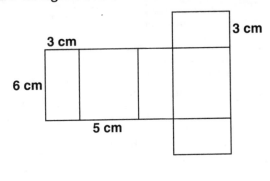

Option 1

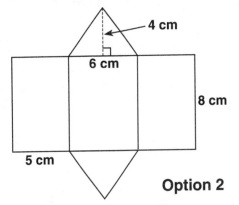

Option 2

Show your work.	**Explain your reasoning.**

Geometry: Answers and Diagnostics

Multiple Choice Questions

Problem #	Correct Answer	Diagnostics
1.	D	A. Student calculated area of entire rectangle. B. Student calculated area of entire rectangle and used incorrect units. C. Student calculated perimeter of rectangle.
2.	C	A. Student chose correct but not *best* description. B. Student assumed all side lengths were the same. D. Student is unclear about the definition of a trapezoid.
3.	D	A. Student multiplied the given dimension by three. B. Student added numerators and denominators instead of multiplying. C. Student added numerators and multiplied denominators.
4.	B	A. Student incorrectly identified the solid formed by the given net. C. Student did not choose the *best* answer. D. Student incorrectly identified the solid formed by the given net.
5.	C	A. Student calculated half of the surface area. B. Student calculated the volume. D. Student incorrectly calculated the lateral area.
6.	C	A. Student multiplied volume by height. B. Student multiplied volume by height and used incorrect units. D. Student used incorrect units.
7.	C	A. Student chose a location that is 11 units from the grocery store only. B. Student chose a location that is 11 units from the post office only. D. Student confused the *x*- and *y*-coordinates.
8.	B	A. Student multiplied 3, 10, and 12, and then divided by 2. C. Student calculated the area of the rectangular portion of the trapezoid. D. Student added all of the numbers together.
9.	B	A. Student multiplied 2 and 4. C. Student multiplied all of the numbers together. D. Student multiplied 2, 3, and 8.
10.	A	B. Student used incorrect units. C. Student calculated volume but included square units. D. Student calculated volume.
11.	A	B. Student used incorrect units. C. Student added dimensions instead of multiplying them. D. Student added dimensions and used incorrect units.
12.	C	A. Student determined length of *RP*. B. Student incorrectly calculated length of *PQ*. D. Student guessed.
13.	D	A. Student multiplied volume by length and width of base. B. Student divided volume by width of base. C. Student divided volume by length of base.
14.	C	A. Student calculated the perimeter of the hexagon. B. Student multiplied 8 and 7. D. Student multiplied 6, 7, and 8.
15.	B	A. Student calculated the area of the rectangle. C. Student calculated the perimeter of the rectangle. D. Student calculated half of the area of the triangle.

Geometry: Answers and Diagnostics

16.	D	A. Student calculated length of *WX*. B. Student guessed. C. Student miscalculated the length of *ZY*.
17.	A	B. Student confused the *x*- and *y*-coordinates. C. Student chose location for *H* that would not yield a parallelogram. D. Student chose location for *H* that would not yield a parallelogram.
18.	A	B. Student multiplied incorrectly. C. Student added area of the base and height. D. Student subtracted height from area of base.
19.	D	A. Student chose a net that forms a rectangular prism. B. Student chose a net that forms a rectangular prism. C. Student chose a net that forms a rectangular prism.
20.	D	A. Student confused surface area and volume. B. Student miscalculated surface area. C. Student confused surface area and volume.

Open-Ended Response Questions

Problem #1

Area of (material for) blue triangle: $\frac{1}{2}$ (12)(8) = 48 in.²; Area of red trapezoid: $\frac{(18 \times 12) - 48}{2}$ = 84 in.²;

Area of white trapezoid: $\frac{(18 \times 12) - 48}{2}$ = 84 in.²

Problem #2

Volume of Box A: $\left(\frac{1}{4}\right)\left(\frac{4}{3}\right)\left(\frac{5}{8}\right)$ = $\frac{5}{24}$ ft³; Volume of Box B: $\left(\frac{3}{5}\right)\left(\frac{7}{8}\right)$ = $\frac{21}{40}$ ft³;
Because Box B holds over twice as much as Box 1 for only $5 more, it is the better deal.

Problem #3

1. (8, –3), (–4, –3), (2, 3), and (2, –9)
2. The point (–4, –3) could be connected to points *A* and *C* to form a parallelogram, because it would be 6 units from point *C*, and make $\overline{AB}$ parallel to $\overline{DC}$ and $\overline{AD}$ parallel to $\overline{BC}$.

Problem #4

Volume of Option 1: 6 x 5 x 3 = 90 cm³;
Surface Area of Option 1: 2(6 x 5) + 2(5 x 3) + 2(3 x 6) = 126 cm²;

Volume of Option 2: $\frac{1}{2}$(6)(4)(8) = 96 cm³;

Surface Area of Option 2: $2\left(\frac{1}{2}(6 \times 4)\right)$ + 2(5 x 8) + (6 x 8) = 152 cm²;

The packages can hold about the same amount. However, Option 2 uses more cardboard. Therefore, Eleesha should choose Option 1.

Ratios and Proportional Relationships

Problem Correlation to CCSS Grade 6 Ratios and Proportional Relationships Standards

MC Problem #	6.RP.A.1	6.RP.A.2	6.RP.A.3	6.RP.A.3a	6.RP.A.3b	6.RP.A.3c	6.RP.A.3d
1					•		
2	•						
3		•					
4						•	
5				•			
6						•	
7							•
8		•					
9			•				
10					•		
11	•						
12						•	
13							•
14		•					
15				•			
16					•		
17	•						
18					•		
19				•			
20							•
Open-Ended Problem #	6.RP.A.1	6.RP.A.2	6.RP.A.3	6.RP.A.3a	6.RP.A.3b	6.RP.A.3c	6.RP.A.3d
1			•				
2	•				•		
3			•			•	
4		•	•	•			

Name: _____   Date: _____

Ratios and Proportional Relationships: Multiple-Choice Assessment Prep

Directions: Circle the choice that best answers the question.

1. A baker can make nine dozen bagels in two hours. How many dozen bagels can he make in eight hours?

   A. 36 dozen

   B. 72 dozen

   C. 432 dozen

   D. 864 dozen

3. A landscaping company paid $96 for 12 shrubs. Which of the following represents the unit rate in this context?

   A. 12:96

   B. 96:12

   C. 1:8

   D. 8:1

2. The ratio of cats to dogs at the local animal shelter is 6:1. Which of the following statements best describes the relationship between the number of cats and dogs at the shelter?

   A. There are one cat and six dogs at the shelter.

   B. There are one dog and six cats at the shelter.

   C. For every cat there are six dogs at the shelter.

   D. For every dog there are six cats at the shelter.

4. Twenty-five percent of the fish in an aquarium are goldfish. There are 124 total fish in the aquarium. How many of the fish are goldfish?

   A. 5

   B. 25

   C. 31

   D. 99

Name: _____ Date: _____

Ratios and Proportional Relationships: Multiple-Choice Assessment Prep

Directions: Circle the choice that best answers the question.

5. The table shows the relationship between cups of blueberries and cups of grapes in a fruit salad recipe. How many cups of grapes would be in a large fruit salad?

Size	Cups of blueberries	Cups of grapes
Small	2	5
Medium	4	10
Large	6	?
Extra large	10	25

A. 21

B. 20

C. 15

D. 12

6. Which of the following expressions can be used to calculate 40% of 205?

A. $\frac{100}{40} \times 205$

B. $\frac{40}{205} \times 100$

C. $\frac{205}{40} \times 100$

D. $\frac{40}{100} \times 205$

7. How many feet are there in 27 yards? (HINT: There are three feet in one yard.)

A. 19,683 ft

B. 81 ft

C. 30 ft

D. 9 ft

8. Chantel can run 13 miles in two hours. Which of the following best describes Chantel's unit running rate for the given units?

A. It takes Chantel about 9.2 minutes to run one mile.

B. It takes Chantel about 0.15 hour to run one mile.

C. Chantel can run 6.5 miles in one hour.

D. Chantel can run 13 miles in two hours.

Name: _____   Date: _____

Ratios and Proportional Relationships: Multiple-Choice Assessment Prep

Directions: Circle the choice that best answers the question.

9. The ratio of the number of petunias to daisies in a flowerbed is 5:4. The total number of flowers in the flowerbed is 180. Use the tape diagram to determine the number of each type of flower.

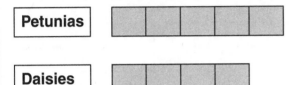

Petunias

Daisies

    A. There are 36 petunias and 45 daisies in the flowerbed.

    B. There are 100 petunias and 80 daisies in the flowerbed.

    C. There are 45 petunias and 36 daisies in the flowerbed.

    D. There are 80 petunias and 100 daisies in the flowerbed.

10. Ami's hair grows at a rate of three inches in four months. At that rate, how much will her hair grow in 12 months?

    A. 12 in.

    B. 9 in.

    C. 6 in.

    D. 3 in.

11. In an orchard, the ratio of peach to apple trees is 3:7. Which of the following statements best describes the relationship between the number of peach trees and apple trees in the orchard?

    A. There are three peach trees and seven apple trees in the orchard.

    B. There are 10 fruit trees in the orchard.

    C. For every three peach trees, there are seven apple trees.

    D. For every three apple trees, there are seven peach trees.

12. Lia cataloged the songs on her MP3 player and realized that 60% of the songs are country. She has 1,710 country songs on her MP3 player. How many total songs does she have?

    A. 684 total songs

    B. 1,026 total songs

    C. 2,850 total songs

    D. 3,420 total songs

Name: _____ Date: _____

Ratios and Proportional Relationships: Multiple-Choice Assessment Prep

Directions: Circle the choice that best answers the question.

13. Speedy Construction can build a new garage every 28 to 31 days. How many garages can they complete in 12 weeks? Assume they finish a garage every 28 days.

   A. 3 garages

   B. 2 garages

   C. 1 garages

   D. 4 garages

15. The data in the table shows the ratio of production of trucks at an automobile factory.

Number of days	Number of trucks
5	180
15	540
?	1,080
50	1,800

How many days would it take to produce 1,080 trucks at this factory?

   A. 25

   B. 30

   C. 35

   D. 40

14. To mix your own potting soil, you need two gallons of compost for every three gallons of peat moss. Which of the following gives the unit rate for compost to peat moss?

   A. $1:\frac{2}{3}$

   B. $\frac{2}{3}:1$

   C. 2:3

   D. 3:2

16. A two-pound box of strawberries costs $5. Harriet needs 10 pounds to make strawberry jam. How much will 10 pounds cost?

   A. $25

   B. $13

   C. $10

   D. $5

Name: _____ Date: _____

Ratios and Proportional Relationships: Multiple-Choice Assessment Prep

Directions: Circle the choice that best answers the question.

---

17. In a recipe for trail mix, the ratio of cups of raisins to total cups of trail mix is 1:5. Which of the following statements best describes the relationship between the number of cups of raisins and the total cups of trail mix?

A. For every cup of raisins, there are a total of four cups of the other ingredients in the trail mix.

B. For every cup of raisins, there are a total of five cups of the other ingredients in the trail mix.

C. For every cup of raisins, there are four cups of each of the other ingredients in the trail mix.

D. For every cup of raisins, there are five cups of each of the other ingredients in the trail mix.

19. The table shows the times and distances for two cyclists.

Cyclist #1

Miles	16	24	32
Time (hours)	2	3	4

Cyclist #2

Miles	18	24	30
Time (hours)	3	4	5

Which of the following statements is true?

A. Both cyclists ride at the same rate.

B. Cyclist #1 is slower than Cyclist #2.

C. Cyclist #2 is faster than Cyclist #1.

D. Cyclist #1 is faster than Cyclist #2.

---

18. An airplane travels 289 miles in two hours. How far can it travel in five hours?

A. 722.5 mi

B. 578 mi

C. 500 mi

D. 115.6 mi

20. Gabriel can run 10 kilometers in 40 minutes. How many miles can he run in 40 minutes? (HINT: 5 kilometers is equal to approximately 3.1 miles.)

A. 31 mi

B. 15.5 mi

C. 12.4 mi

D. 6.2 mi

Name: _____ Date: _____

Ratios and Proportional Relationships: Open-Ended Response Assessment Prep

Directions: Answer the question completely. Show your work and explain your reasoning.

Problem 1: The jazz band sells 375 tickets for their concert. The ratio of the number of child tickets sold to the number of adult tickets sold is 2:3. Draw a tape diagram to model this situation. Then determine the number of each type of ticket that is sold.

Show your work.	Explain your reasoning.

Name: _____   Date: _____

Ratios and Proportional Relationships:
Open-Ended Response Assessment Prep

Directions: Answer the question completely. Show your work and explain your reasoning.

Problem 2: There are 32 chickens and 12 sheep on a farm. The ratio of sheep to horses is 4:1.

1. Write a ratio expressing the relationship between the chicken and the sheep. Then explain what that ratio means.

2. Determine the number of horses on the farm.

Show your work.	**Explain your reasoning.**

Name: _____ Date: _____

Ratios and Proportional Relationships: Open-Ended Response Assessment Prep

Directions: Answer the question completely. Show your work and explain your reasoning.

Problem 3: About 17% of Americans have blue eyes. There are 85 students at Park Street Middle School who have blue eyes. How many total students would there need to be at Park Street Middle School for the percentage of students with blue eyes to match the national statistic?

Show your work.	Explain your reasoning.

Name: _____ Date: _____

Ratios and Proportional Relationships: Open-Ended Response Assessment Prep

Directions: Answer the question completely. Show your work and explain your reasoning.

Problem 4: The tables below show the relationship between time and distance walked by two different people in a walk-a-thon. Complete the tables and determine which person walked at a faster rate. Graph the ordered pairs, creating a line for each walker, and use the graph to help support your answer.

Walker #1

Time (hours)	1	2	3	4	5
Distance (miles)		6		12	15

Walker #2

Time (hours)	2	4		8	10
Distance (miles)	5		15	20	

Show your work.

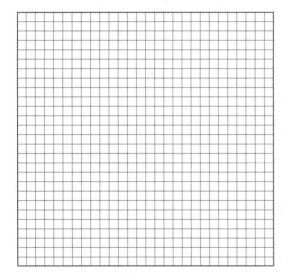

Explain your reasoning.

Ratios and Proportional Relationships:
Answers and Diagnostics

Multiple-Choice Questions

Problem #	Correct Answer	Diagnostics
1.	A	B. Student multiplied 9 by 8 instead of 4. C. Student calculated the actual number of bagels. D. Student multiplied 9 by 12 and then by 8.
2.	D	A. Student confused dogs and cats and incorrectly interpreted what the ratio means. B. Student incorrectly interpreted what the ratio means. This answer could be true, but it is not the *best* answer. C. Student confused dogs and cats.
3.	D	A. Student chose the ratio of total shrubs to total amount paid. B. Student chose the ratio of total amount paid to total shrubs. C. Student chose the ratio of one shrub to amount paid for one shrub.
4.	C	A. Student divided 124 by 25 and then rounded to the nearest whole number. B. Student chose the percent of goldfish in the aquarium. D. Student subtracted 25 from 124.
5.	C	A. Student subtracted 4 from 25. B. Student doubled 10 or subtracted 5 from 25. D. Student added 2 to 10.
6.	D	A. Student represented the ratio for 40% incorrectly. B. Student chose the expression that could be used to determine what percent 40 is of 205. C. Student chose the expression that could be used to determine what percent 205 is of 40.
7.	B	A. Student cubed 27. C. Student added 3 to 27. D. Student divided 27 by 3.
8.	C	A. Student chose an answer that is true but is not the unit rate for the number of miles Chantel can run in one hour. B. Student chose an answer that is true but is not the unit rate for the number of miles Chantel can run in one hour. D. Student chose an answer that is true but is not the unit rate for the number of miles Chantel can run in one hour.
9.	B	A. Student divided 180 by 5 and 4. C. Student divided 180 by 5 and 4 and then confused petunias and daisies. D. Student confused petunias and daisies.
10.	B	A. Student miscalculated the rate at which Ami's hair grows. C. Student calculated the amount Ami's hair would grow in 8 months. D. Student chose the number of inches Ami's hair grows in 4 months.
11.	C	A. Student incorrectly interpreted what the ratio means. This answer could be true, but it is not the *best* answer. B. Student incorrectly interpreted what the ratio means. This answer could be true, but it is not the *best* answer. D. Student confused peach and apple trees.

Ratios and Proportional Relationships: Answers and Diagnostics

12.	C	A. Student calculated 40% of 1,710. B. Student calculated 60% of 1,710. D. Student doubled 1,710.
13.	A	B. Student represented one week with five days. C. Student guessed. D. Student forgot to convert 12 weeks to 84 days.
14.	B	A. Student chose ratio of peat moss to compost. C. Student chose ratio of compost to peat moss, but it is not the unit rate. D. Student chose ratio of peat moss to compost, but it is not the unit rate.
15.	B	A. Student chose the number of days that is 10 more than 15. C. Student chose the number that is about halfway between 15 and 50. D. Student chose the number that is 10 less than 50.
16.	A	B. Student noticed that 2 plus 8 is 10, so they added 8 to $5. C. Student miscalculated the unit rate. D. Student guessed.
17.	A	B. Student failed to recognize that the cup of raisins is included in the five cups of trail mix. C. Student incorrectly interpreted the five in the ratio as meaning the amount of each of the other ingredients once the cup of raisins was subtracted from the total. D. Student incorrectly interpreted the five in the ratio as meaning the amount of each of the other ingredients.
18.	A	B. Student doubled 289. C. Student guessed. D. Student calculated 40% of 289.
19.	D	A. Student looked at the second distance listed in both tables and assumed it meant that the cyclists traveled at the same rate. B. Student did not understand how to compare the cycling rates. C. Student did not understand how to compare the cycling rates.
20.	D	A. Student used the rate of 1 km to 3.1 mi. B. Student multiplied 5 by 3.1. C. Student multiplied 4 by 3.1.

Open-Ended Response Questions
Problem #1

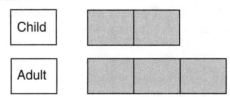

Because there are five parts, one part represents 375 ÷ 5 or 75. Therefore, there were 2 x 75 or 150 child tickets sold and 3 x 75 or 225 adult tickets sold.

Problem #2
1. 8:3; There are eight chickens for every three sheep.
2. There are three horses on the farm because the ratio 12:3 is equivalent to 4:1.

Ratios and Proportional Relationships: Answers and Diagnostics

Problem #3

$$\frac{85}{?} = \frac{17}{100} \longrightarrow (85)(100) = 17x \longrightarrow 8{,}500 \div 17 = 500$$

There would need to be 500 total students at Park Street Middle School.

Problem #4

Walker #1

Time (hours)	1	2	3	4	5
Distance (miles)	3	6	9	12	15

Walker #2

Time (hours)	2	4	6	8	10
Distance (miles)	5	10	15	20	25

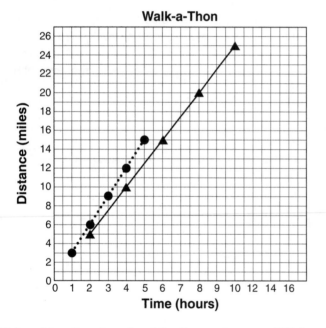

● Walker #1

▲ Walker #2

Walker #1 walks at a rate of 3 miles per hour, and Walker #2 walks at a rate of 2.5 miles per hour. So, Walker #1 walks at a faster rate. This can also be seen in the graph where the points create a line with a steeper slope for Walker #1 than for Walker #2.

The Number System

Problem Correlation to CCSS Grade 6 The Number System Standards

MC Problem #	6.NS.A.1	6.NS.B.2	6.NS.B.3	6.NS.B.4	6.NS.C.5	6.NS.C.6a	6.NS.C.6b	6.NS.C.6c	6.NS.C.7a	6.NS.C.7b	6.NS.C.7c	6.NS.C.7d	6.NS.C.8
1	•												
2	.	•											
3				•									
4						•							
5							•						
6					•								
7									•				
8											•		
9			•										
10	•												
11						•							
12				•									
13										•			
14					•								
15									•				
16											•		
17								•					
18										•			
19			•										
20										•			
21												•	
22			•										
23													•
24			•										

Open-Ended Problem #	6.NS.A.1	6.NS.B.2	6.NS.B.3	6.NS.B.4	6.NS.C.5	6.NS.C.6a	6.NS.C.6b	6.NS.C.6c	6.NS.C.7a	6.NS.C.7b	6.NS.C.7c	6.NS.C.7d	6.NS.C.8
1			•										
2	•												
3						•		•	•	•	•		•

Name: _____ Date: _____

The Number System: Multiple-Choice Assessment Prep

Directions: Circle the choice that best answers the question.

---

1. What is the simplified form of the quotient of $\frac{3}{5}$ and $\frac{15}{16}$?

    A. $\frac{16}{25}$

    B. $\frac{9}{16}$

    C. $\frac{45}{80}$

    D. $\frac{48}{75}$

4. Which of the following is not equal to 7?

    A. $-(7)$

    B. $-(-7)$

    C. $|-7|$

    D. $-(-|-7|)$

---

2. Determine the quotient of 2,944 and 23.

    A. 67,712

    B. 1,280

    C. 150

    D. 128

5. In which quadrant is the point $(-3, -6)$ located?

    A. Quadrant I

    B. Quadrant II

    C. Quadrant III

    D. Quadrant IV

---

3. What is the greatest common factor of 36 and 48?

    A. 144

    B. 12

    C. 6

    D. 2

6. Which number best represents a temperature of 8 degrees below 0?

    A. 8

    B. $-(-8)$

    C. -8

    D. $|-8|$

Name: _____   Date: _____

The Number System: Multiple-Choice Assessment Prep

Directions: Circle the choice that best answers the question.

---

7. Which point represents $\frac{9}{5}$?

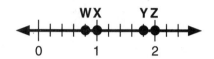

A. Z

B. Y

C. X

D. W

---

10. Pedro has a piece of wood that is $\frac{8}{9}$ yard long. He cuts it into pieces that are $\frac{1}{3}$ yard long. How many $\frac{1}{3}$-yard pieces does Pedro have?

A. 0

B. 1

C. 2

D. 3

---

8. Is the following inequality true?

$$|-14| > |7|$$

A. Yes, because −14 is further from 0 on the number line than 7.

B. No, because −14 is further from 0 on the number line than 7.

C. Yes, because −14 is less than 7.

D. No, because −14 is less than 7.

---

11. Which point on the number line is the opposite of 5?

A. A

B. B

C. C

D. D

---

9. Calculate the sum of 97.24 and 324.881.

A. 322.121

B. 334.605

C. 422.121

D. 31,591.42844

---

12. What is the least common multiple of 9 and 12?

A. 3

B. 18

C. 24

D. 36

Name: _____ Date: _____

The Number System: Multiple-Choice Assessment Prep

Directions: Circle the choice that best answers the question.

---

13. Which statement does not describe the relationship shown in the inequality?

$$-9 < -2$$

A. Negative 9 is to the left of –2 on the number line.

B. Negative 2 is to the right of –9 on the number line.

C. Negative 9 is to the right of –2 on the number line.

D. Negative 9 is further from 0 than –2 on the number line.

15. Which point is located at $\left(-4, \frac{3}{2}\right)$?

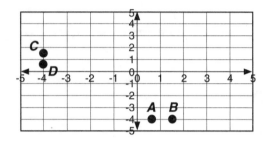

A. A

B. B

C. C

D. D

---

14. Jacques went hiking in a canyon. He started at an elevation of 948 feet. At the end of his hike, he was at an elevation of –948 feet. What does 0 feet represent in this scenario?

A. The change in Jacques' elevation

B. Sea level

C. The distance Jacques hiked

D. Cannot be determined

16. When Barbara got up in the morning, the temperature was 0°C. By 8:00 P.M., the temperature had dropped to –9°C. Which expression best represents the magnitude of change in the temperature?

A. $-9 < 9$

B. $-9 < 0$

C. $|-9| > 0$

D. $|-9| = 9$

Name: _____ Date: _____

The Number System: Multiple-Choice Assessment Prep

Directions: Circle the choice that best answers the question.

17. Point A is located at (5, 9) and point B is located at (−5, −9). Which statement best describes the relationship between the points?

A. Point B is the reflection of point A across the x-axis.

B. Point B is the reflection of point A across the y-axis.

C. Point B is the reflection of point A across the x-axis and the y-axis.

D. None of the above

19. Calculate the quotient.

$$31.18 \overline{)85.5891}$$

A. 0.002745

B. 0.02745

C. 2.745

D. 27.45

18. Caterina is scuba diving. After diving for a time, she pauses to watch a school of fish. The inequality represents the relationship between Caterina's depth at the beginning of her dive and when she stops to watch the fish:

$$-10ft > -40ft$$

Which of the following statements is true?

A. Caterina is closer to the surface of the water at the beginning of her dive.

B. Caterina is closer to the surface of the water when she stops to watch the fish.

C. Caterina is further from the surface of the water at the beginning of her dive.

D. None of the above

20. The temperature on Monday was −2°F. The temperature on Tuesday was −5°F. The relationship between these temperatures can be represented by the inequality:

$$-2°F > -5°F$$

Which of the following statements is true?

A. It was warmer on Tuesday than on Monday.

B. It was warmer on Monday than on Tuesday.

C. It was colder on Monday than on Tuesday.

D. None of the above

Name: _____ Date: _____

The Number System: Multiple-Choice Assessment Prep

Directions: Circle the choice that best answers the question.

---

21. Which of the following statements is true?

   A. An account with a balance that is less than –$20 indicates a debt that is less than $20.

   B. An account with a balance that is greater than –$20 indicates a debt that is greater than $20.

   C. An account with a balance that is less than –$20 indicates no debt.

   D. An account with a balance that is less than –$20 indicates a debt that is greater than $20.

---

23. Which expression can be used to determine the distance between points *A* and *B*?

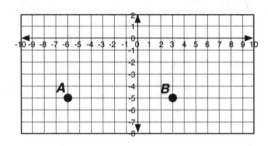

   A. $|-6 + 3| = 3$

   B. $|-6 - 3| = 9$

   C. $-6 - 3 = -9$

   D. $|-5 - (-5)| = 0$

---

22. Calculate the difference.

$$409.31 - 186.2$$

   A. 133.11

   B. 223.11

   C. 233.11

   D. 3,906.9

---

24. Determine the product of 86.234 and 29.4.

   A. 2,535.2796

   B. 253.52796

   C. 115.634

   D. 2.933

Name: _____ Date: _____

The Number System:
Open-Ended Response Assessment Prep

Directions: Answer the question completely. Show your work and explain your reasoning.

Problem 1: Mr. Erb asks his students to determine the quotient of 693.778 and 39.5. The work of one of his students is shown. Identify the mistake the student made. Then correct it and determine the actual quotient.

```
              1.7564
       39.5 ) 693.7780
             −395
              2987
             −2765
              2227
             −1975
              2528
             −2370
              1580
             −1580
                 0
```

Show your work.

Explain your reasoning.

Name: _____ Date: _____

The Number System:
Open-Ended Response Assessment Prep

Directions: Answer the question completely. Show your work and explain your reasoning.

Problem 2: Write a story context for the division expression shown below. Then draw a fraction model to show the quotient.

$$\frac{8}{9} \div \frac{1}{6}$$

Show your work.

Explain your reasoning.

Name: _____ Date: _____

The Number System:
Open-Ended Response Assessment Prep

Directions: Answer the question completely. Show your work and explain your reasoning.

Problem 3: Your teacher asks you to plot the following numbers on a number line.

$$0.2, \quad \frac{4}{5}, \quad -\frac{9}{10}, \quad -0.7, \quad \frac{7}{10}$$

1. Would number line #1 or number line #2 allow you to more accurately plot the numbers? Explain your reasoning.

Number line #1

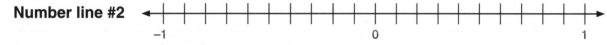

Number line #2

2. Plot and label the given numbers on the number line you chose in part 1.

3. Write an inequality statement that correctly compares the numbers.

4. Which numbers have the same absolute value? Explain your reasoning.

Show your work.	**Explain your reasoning.**

The Number System: Answers and Diagnostics

Multiple-Choice Questions

Problem #	Correct Answer	Diagnostics
1.	A	B. Student chose the product of $\frac{3}{5}$ and $\frac{15}{16}$. C. Student chose the non-simplified product of $\frac{3}{5}$ and $\frac{15}{16}$. D. Student chose the non-simplified quotient of $\frac{3}{5}$ and $\frac{15}{16}$.
2.	D	A. Student calculated the product of 23 and 2,944. B. Student misplaced the decimal point in the quotient. C. Student estimated the quotient.
3.	B	A. Student chose the least common multiple. C. Student chose a common factor that is not the *greatest* common factor. D. Student chose a common factor that is not the *greatest* common factor.
4.	A	B. Student did not understand that the opposite of the opposite of a number is the number itself. C. Student did not understand that the absolute value of a number measures the distance the number is from 0 and is always positive. D. Student did not understand that the absolute value of a number measures the distance the number is from 0 and is always positive or that the opposite of the opposite of a number is the number itself.
5.	C	A. Student did not understand how the signs of the numbers are related to the location of a point in the coordinate plane. B. Student did not understand how the signs of the numbers are related to the location of a point in the coordinate plane. D. Student did not understand how the signs of the numbers are related to the location of a point in the coordinate plane.
6.	C	A. Student chose the number that would best represent 8 degrees above 0. B. Student chose a number that has a positive value and could represent 8 degrees above 0. D. Student chose a number that has a positive value and could represent 8 degrees above 0.
7.	B	A. Student chose the point at 2, but $\frac{9}{5}$ is equal to $1\frac{4}{5}$. C. Student chose the point at 1, but $\frac{9}{5}$ is equal to $1\frac{4}{5}$. D. Student chose the point that represents the fractional part when $\frac{9}{5}$ is converted into a mixed number.
8.	A	B. Student did not understand what absolute value means. C. Student did not understand what absolute value means. D. Student did not understand what absolute value means.
9.	C	A. Student regrouped incorrectly when adding digits in hundreds place. B. Student failed to line up decimal points before adding. D. Student calculated the product.
10.	C	A. Student multiplied the fractions, resulting in a product less than 1. B. Student guessed. D. Student rounded $2\frac{2}{3}$ to 3.

The Number System: Answers and Diagnostics

11.	A	B. Student did not understand how to locate the opposite of a number on a number line. C. Student chose the point that represents 5 instead of its opposite. D. Student did not understand how to locate the opposite of a number on a number line.
12.	D	A. Student chose the greatest common factor. B. Student chose a number that is a multiple of 9 but not of 12. C. Student chose a number that is a multiple of 12 but not of 9.
13.	C	A. Student did not understand how to order integers. B. Student did not understand how to order integers. D. Student did not understand how integers are represented on a number line.
14.	B	A. Student did not understand what positive and negative elevations represent. C. Student did not understand what positive and negative elevations represent. D. Student guessed.
15.	C	A. Student confused the x- and y-coordinates and did not understand that $\frac{3}{2}$ is greater than 1. B. Student confused the x- and y-coordinates. D. Student did not understand that $\frac{3}{2}$ is greater than 1.
16.	D	A. Student did not understand what the question was asking. B. Student chose a true expression, but it did not answer the question. C. Student chose a true expression, but it did not answer the question.
17.	C	A. Student did not understand that reflecting a point across the x-axis changes only the sign of the y-coordinate. B. Student did not understand that reflecting a point across the y-axis changes only the sign of the x-coordinate. D. Student did not understand that reflecting a point across *both* axes changes the sign of both coordinates.
18.	A	B. Student did not understand how negative depths relate. C. Student did not understand how negative depths relate. D. Student guessed.
19.	C	A. Student misplaced the decimal point in the quotient. B. Student misplaced the decimal point in the quotient. D. Student misplaced the decimal point in the quotient.
20.	B	A. Student did not understand how negative temperatures relate. C. Student did not understand how negative temperatures relate. D. Student guessed.
21.	D	A. Student did not understand what a negative account balance means. B. Student did not understand what a negative account balance means. C. Student did not understand what a negative account balance means.
22.	B	A. Student regrouped incorrectly. C. Student regrouped incorrectly. D. Student did not align decimal points before subtracting.
23.	B	A. Student added the x-coordinates. C. Student forgot to take the absolute value because distance should be positive. D. Student subtracted the y-coordinates.

The Number System: Answers and Diagnostics

24.	A	B. Student misplaced the decimal point in the product. C. Student added numbers. D. Student divided numbers.

Open-Ended Response Questions

Problem #1

The student forgot to move the decimal place one place to the right before dividing. The correct quotient is 17.564.

Problem #2

Sample answer. Angelique has a piece of ribbon that is $\frac{8}{9}$ yard long. She plans to cut it into $\frac{1}{6}$-yard pieces. How many $\frac{1}{6}$-yard pieces will she have?

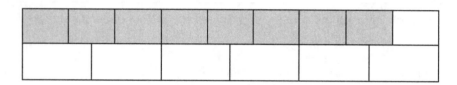

Angelique will have five $\frac{1}{6}$-yard pieces.

Problem #3

1. Sample answer: Number line #2 will allow you to plot the numbers more accurately because it is divided into tenths. All of the given numbers can be written with the common denominator of 10.

2.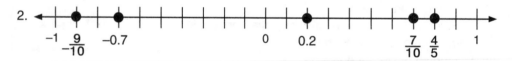

3. $-\frac{9}{10} < -0.7 < 0.2 < \frac{7}{10} < \frac{4}{5}$

4. The numbers −0.7 and $\frac{7}{10}$ have the same absolute value because they are the same distance from 0.

Expressions and Equations

Problem Correlation to CCSS Grade 6 Expressions and Equations Standards

MC Problem #	6.EE.A.1	6.EE.A.2a	6.EE.A.2b	6.EE.A.2c	6.EE.A.3	6.EE.A.4	6.EE.B.5	6.EE.B.6	6.EE.B.7	6.EE.B.8	6.EE.C.9
1			•								
2							•				
3	•										
4						•					
5					•						
6									•		
7	•										
8								•			
9				•							
10		•									
11							•				
12				•							
13				•							
14							•				
15									•		
16		•									
17											•
18								•			
19			•								
20										•	
21			•								
22					•						
Open-Ended Problem #	6.EE.A.1	6.EE.A.2a	6.EE.A.2b	6.EE.A.2c	6.EE.A.3	6.EE.A.4	6.EE.B.5	6.EE.B.6	6.EE.B.7	6.EE.B.8	6.EE.C.9
1				•			•				•
2								•		•	
3							•	•	•		

Name: _____ Date: _____

Expressions and Equations:
Multiple-Choice Assessment Prep

Directions: Circle the choice that best answers the question.

---

1. Which part of the expression could be described as a coefficient?

$$(9 - p)^2 + 5p$$

   A. p

   B. 2

   C. 5

   D. 9

3. Evaluate.

$$6^2 + 9 \div 3 - (1 + 2)^3$$

   A. 12

   B. 30

   C. −12

   D. −20

---

2. Which value of x makes the equation true?

$$8x = 24$$

   A. $x = 192$

   B. $x = 16$

   C. $x = 4$

   D. $x = 3$

4. Which statement best explains why the expression $7x - 2x$ is equivalent to $5x$?

   A. The expressions are equivalent because x is equal to 5.

   B. The expressions are equivalent because they name the same number regardless of which number x stands for.

   C. The expressions are equivalent because x is equal to 0.

   D. None of the above because $7x - 2x$ is not equivalent to $5x$.

Name: _____　Date: _____

Expressions and Equations:
Multiple-Choice Assessment Prep

Directions: Circle the choice that best answers the question.

---

5. Which expression is equivalent to $6(x - 3)$?

　A. $6x - 3$

　B. $6x - 18$

　C. $3x$

　D. $6x + 18$

---

8. Jin has 17 more posters on her wall than her sister has on her wall. Which expression represents the number of posters Jin has, if the number of posters her sister has is represented by p?

　A. $17p$

　B. $17 - p$

　C. $p - 17$

　D. $p + 17$

---

6. Kiana is four times as old as Noah. Kiana is 32. Which equation can be used to determine Noah's age?

　A. $4 + n = 32$

　B. $4n = 32$

　C. $\dfrac{n}{4} = 32$

　D. $32n = 4$

---

9. Evaluate the given expression when b is equal to six.

$$4b^2 - 9$$

　A. 567

　B. 135

　C. 108

　D. 16

---

7. Which expression represents the sum of 3 and 5 squared divided by the difference of 11 and 7?

　A. $3^2 + 5^2 \div (11 - 7)$

　B. $(3 + 5)^2 \div 11 - 7$

　C. $3^2 + 5^2 \div 11 - 7$

　D. $(3 + 5)^2 \div (11 - 7)$

---

10. Choose the expression that represents the difference of 3 times a number and 15.

　A. $3(n - 15)$

　B. $3n - 15$

　C. $15 - 3n$

　D. $3n + 15$

Name: _____ Date: _____

Expressions and Equations:
Multiple-Choice Assessment Prep

Directions: Circle the choice that best answers the question.

---

11. Solve.

$$4g \leq 28$$

A. $g \leq 7$

B. $g < 7$

C. $g \geq 7$

D. $g = 7$

---

14. Which value(s) of w make the given inequality true?

$$32 - w > 5$$

A. $w = 16$

B. $w = 10$

C. $w = 0$

D. All of the above

---

12. Which expression is equivalent to $2x + 4(3x + 2y) - 3y$?

A. $14x - y$

B. $14x + 11y$

C. $14x + 5y$

D. $18x^2 + 12xy - 3y$

---

15. Ronna has 13 more chickens on her farm than Perry has on his. Ronna has 49 chickens on her farm. Which equation can be used to determine the number of chickens Perry has?

A. $13p = 49$

B. $p - 13 = 49$

C. $p + 49 = 13$

D. $p + 13 = 49$

---

13. Which expression is *not* equivalent to $72p - 48q$?

A. $24(3p + 2q)$

B. $24(3p - 2q)$

C. $8(9p - 6q)$

D. $-8(-9p + 6q)$

---

16. Choose the expression that represents the quotient of eight and x.

A. $x \div 8$

B. $8 \div x$

C. $8x$

D. $8 - x$

Name: _____   Date: _____

Expressions and Equations:
Multiple-Choice Assessment Prep

Directions: Circle the choice that best answers the question.

---

17. Rei sells paper cranes for $2 each. Let n represent Rei's income and c represent the number of cranes Rei sells. Which equation could be used to determine Rei's profits for a given number of paper cranes?

  A. $c = 2 + n$

  B. $n = 2 + c$

  C. $c = 2n$

  D. $n = 2c$

---

20. Leon runs more than 20 miles each week. Which expression can be used to represent the number of miles Leon runs each week?

  A. $m \geq 20$

  B. $m > 20$

  C. $m < 20$

  D. $m = 20$

---

18. What does b represent in the given inequality?

$$b \leq 26$$

  A. Any number

  B. Any number that is less than 26

  C. Any number that is less than or equal to 26

  D. 26

---

21. Which word best describes $5y^4$ as it appears in the given expression?

$$5y^4 + 28$$

  A. Constant

  B. Coefficient

  C. Sum

  D. Term

---

19. The expression $16t^2$ can be used to determine the distance, in feet, a falling object travels for a given time, t, in seconds. How many feet will a falling object travel in 3 seconds?

  A. 2,304 ft

  B. 144 ft

  C. 96 ft

  D. 48 ft

---

22. Which expression is equivalent to $36a - 54b$?

  A. $18(2a - 4b)$

  B. $18(2a - 54b)$

  C. $18(2a + 3b)$

  D. $18(2a - 3b)$

Name: _____ Date: _____

Expressions and Equations:
Open-Ended Response Assessment Prep

Directions: Answer the question completely. Show your work and explain your reasoning.

Problem 1: Eggs are a good source of protein and vitamins A, E, and K. One egg has 6 grams of protein.

1. Write an equation that expresses the relationship between the number of eggs, *e*, and the total amount of protein, *p*.

2. Identify the independent and dependent variables in your equation.

3. Use your equation to complete the table.

4. Graph the ordered pairs from your table. Be sure to label the axes.

Show your work.

Explain your reasoning.

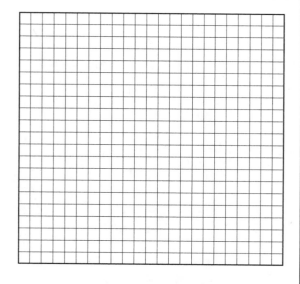

Eggs	1	2	5	9
Protein (grams)				

Name: _____ Date: _____

Expressions and Equations:
Open-Ended Response Assessment Prep

Directions: Answer the question completely. Show your work and explain your reasoning.

Problem 2: Wan covers her rose bushes when the temperature drops below 27°F.

1. Write an inequality to represent the temperatures at which Wan covers her rose bushes.

2. Graph the solutions to the inequality on the number line.

Show your work.	**Explain your reasoning.**

Name: _____ Date: _____

Expressions and Equations:
Open-Ended Response Assessment Prep

Directions: Answer the question completely. Show your work and explain your reasoning.

Problem 3: Tonya says that her favorite number is 44. Pena says that Tonya's number is 9 more than her favorite number.

1. Write an equation to represent the relationship between Tonya's and Pena's favorite numbers.

2. Solve the equation to determine Pena's favorite number.

Show your work.	Explain your reasoning.

Expressions and Equations: Answers and Diagnostics

Multiple-Choice Questions

Problem #	Correct Answer	Diagnostics
1.	C	A. Student chose the variable. B. Student chose the exponent. D. Student chose the constant.
2.	D	A. Student multiplied 24 by 8. B. Student subtracted 8 from 24. C. Student incorrectly divided 24 by 8.
3.	A	B. Student cubed each of the numbers in the parentheses. C. Student added before dividing by 3. D. Student multiplied 6 by 2 and then added 9 before dividing by 3.
4.	B	A. Student did not understand that x represented a variable in the expressions. C. Student did not understand that x represented a variable in the expressions. D. Student did not understand how to determine whether or not expressions are equivalent.
5.	B	A. Student did not distribute the 6 to both the x and the 3. C. Student multiplied only the x by 6 and then subtracted 3 as if they were like terms. D. Student made a sign error when distributing the 6.
6.	B	A. Student did not understand that *four times* indicates multiplication. C. Student did not understand that *four times* indicates multiplication. D. Student guessed.
7.	D	A. Student did not understand what *the sum of 3 and 5 squared* means. B. Student did not understand what *divided by the difference of 11 and 7* means. C. Student did not understand how to translate words into an expression.
8.	D	A. Student did not understand that *more than* indicates addition. B. Student did not understand that *more than* indicates addition. C. Student did not understand that *more than* indicates addition.
9.	B	A. Student multiplied 6 by 4 before squaring. C. Student subtracted 9 from b^2 before multiplying by 4. D. Student divided by 9 rather than subtracting.
10.	B	A. Student chose the expression that represents 3 times the difference of a number and 15. C. Student chose the expression that represents the difference of 15 and 3 times a number. D. Student chose the expression that represents the sum of 3 times a number and 15.
11.	A	B. Student disregarded that the quantities could be equivalent. C. Student reversed the inequality symbol. D. Student disregarded that $4g$ could be less than 28.
12.	C	A. Student did not distribute the 4 to the $2y$. B. Student added $8y$ and $3y$ instead of subtracting them. D. Student incorrectly added the unlike terms $2x$ and 4 before distributing.
13.	A	B. Student chose an expression that is equivalent to $72p - 48q$. C. Student chose an expression that is equivalent to $72p - 48q$. D. Student chose an expression that is equivalent to $72p - 48q$.

Expressions and Equations: Answers and Diagnostics

14.	D	A. Student chose only one of the values that makes the inequality true. B. Student chose only one of the values that makes the inequality true. C. Student chose only one of the values that makes the inequality true.
15.	D	A. Student did not understand that *more than* indicates addition. B. Student did not understand that *more than* indicates addition. C. Student misunderstood what each of the numbers in the problem represented.
16.	B	A. Student chose the expression that represents the quotient of *x* and eight. C. Student chose the expression that represents the product of eight and *x*. D. Student chose the expression that represents the difference of eight and *x*.
17.	D	A. Student guessed. B. Student did not understand that *$2 each* indicates multiplication. C. Student did not understand the relationship between the variable quantities.
18.	C	A. Student did not understand the meaning of the inequality symbol. B. Student did not understand the meaning of the inequality symbol. D. Student did not understand the meaning of the inequality symbol.
19.	B	A. Student multiplied by 16 before squaring. C. Student multiplied three by two rather than squaring it. D. Student disregarded the exponent.
20.	B	A. Student chose the symbol indicating greater than *or equal to* 20. C. Student chose the symbol indicating less than 20. D. Student did not understand that *more than* indicated an inequality symbol rather than an equal sign.
21.	D	A. Student chose the word that best describes the 28 in the expression. B. Student chose the word that best describes the 5 in the expression. C. Student chose the word that best describes the entire expression.
22.	D	A. Student incorrectly factored 18 out of 54. B. Student did not factor 18 out of 54. C. Student made a sign error.

Open-Ended Response Questions

Problem #1

1. $p = 6e$
2. The amount of protein depends on the number of eggs. So, *p* is the dependent variable, and *e* is the independent variable.
3.

Eggs	1	2	5	9
Protein (grams)	6	12	30	54

4.

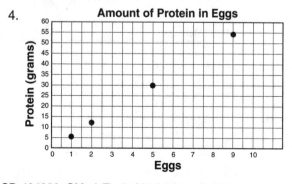

Problem #2

1. Let *t* represent temperature, then $t < 27$ represents the temperatures at which Wan covers her rose bushes.
2.

Problem #3

1. Let *p* represent Pena's favorite number, then $p + 9 = 44$.
2. $p + 9 = 44 \longrightarrow p + 9 - 9 = 44 - 9 \longrightarrow$ $p = 35$; Pena's favorite number is 35.

Statistics and Probability

Problem Correlation to CCSS Grade 6 Statistics and Probability Standards

MC Problem #	6.SP.A.1	6.SP.A.2	6.SP.A.3	6.SP.B.4	6.SP.B.5a	6.SP.B.5b	6.SP.B.5c	6.SP.B.5d
1			•					
2					•			
3				•				
4	•							
5				•				
6			•					
7								•
8							•	
9	•							
10								•
11							•	
12					•		•	
13							•	
14					•			
15				•				
16								•
17						•		
18		•						
19				•				
20							•	

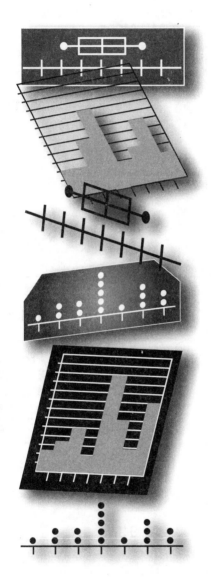

Open-Ended Problem #	6.SP.A.1	6.SP.A.2	6.SP.A.3	6.SP.B.4	6.SP.B.5a	6.SP.B.5b	6.SP.B.5c	6.SP.B.5d
1	•			•			•	
2			•	•			•	•
3		•		•				•

Name: _____ Date: _____

Statistics and Probability:
Multiple-Choice Assessment Prep

Directions: Circle the choice that best answers the question.

1. Which one best describes *measure of center* for a numerical data set?

   A. Mean

   B. Median

   C. A single number that summarizes all of the values

   D. How the data values vary with a single number

3. Which one *cannot* be used to display numerical data on a number line?

   A. Histogram

   B. Dot plot

   C. Bar graph

   D. Box-and-whisker plot

2. The frequency chart shows the results of a survey Jesse conducted.

Eye color	Frequency
Blue	23
Brown	41
Green	8
Other	19

   How many people did Jesse survey?

   A. 91

   B. 72

   C. 19

   D. 4

4. Is the given question an example of a statistical question?

   What is your favorite color?

   A. No, because it does not anticipate or account for variability in the answers.

   B. No, because it anticipates and accounts for variability in the answers.

   C. Yes, because it does not anticipate or account for variability in the answers.

   D. Yes, because it anticipates and accounts for variability in the answers.

Name: _____ Date: _____

Statistics and Probability:
Multiple-Choice Assessment Prep

Directions: Circle the choice that best answers the question.

5. The graph shows the quiz scores for Mr. Becker's math class.

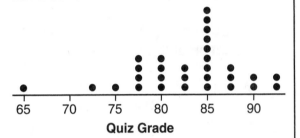

Mr. Becker's Math Class Quiz Grades

Quiz Grade

What type of graph is shown?

A. Box-and-whisker plot

B. Histogram

C. Dot plot

D. None of the above

6. Mrs. Garrett asks each of her students to record the number of hours they spend reading in one week. After she collects this data, she would like to summarize it with a single number. Which of the following would do that?

A. Mean absolute deviation

B. Measures of center

C. Interquartile range

D. Measures of variability

7. The number of copies of a new mystery novel sold at the local bookstore each day for a week is listed below.

1, 8, 10, 15, 16, 17, 48

Which measure of center would best describe the average daily sales of the new mystery novel?

A. Median, because there is an outlier.

B. Median, because there is *not* an outlier.

C. Mean, because there is an outlier.

D. Mean, because there is *not* an outlier.

8. Blane's scores on his last seven golf games are listed. What is Blane's mean golf score?

109, 99, 95, 97, 101, 101, 98

A. 99

B. 100

C. 101

D. 700

Name: _____ Date: _____

Statistics and Probability:
Multiple-Choice Assessment Prep

Directions: Circle the choice that best answers the question.

9. Aiden would like to collect some data about the students in his school. He plans to make a survey that includes several questions. Help Aiden by identifying which of these questions is a good example of a statistical question.

   A. How tall are you?

   B. How many cousins do you have?

   C. How many pets do you have?

   D. All of the above

11. The foot lengths, in centimeters, of the boys in Mrs. Fielding's class are listed.

   22, 23, 23, 23, 23, 24, 24, 25, 25, 26, 26

   What is the approximate mean absolute deviation of this data set?

   A. 24

   B. 4

   C. 1.09

   D. None of the above

10. The box-and-whisker plot shows the distribution of heights of the members of the basketball team.

   Heights of Players on Basketball Team

   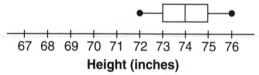

   Height (inches)

   Which measure of center and variability would best describe the distribution of the data?

   A. Mean and interquartile range

   B. Mean and mean absolute deviation

   C. Median and interquartile range

   D. Median and mean absolute deviation

12. Kirby surveyed his friends about the number of cousins they have. The results are shown in the dot plot. Which statement about the data is *not* true?

   Kirby's Friends' Cousin Count

   Number of Cousins

   A. The mean number of cousins that Kirby's friends have is 6.

   B. Most of Kirby's friends have between 2 and 10 cousins.

   C. A few of Kirby's friends have more than 20 cousins.

   D. Kirby surveyed 28 of his friends.

Name: _____ Date: _____

Statistics and Probability:
Multiple-Choice Assessment Prep

Directions: Circle the choice that best answers the question.

---

13. The number of points the basketball team scored in each of their first five games is listed.

69, 66, 81, 77, 72

What is the median number of points scored in the first five games?

A. 72

B. 73

C. 79

D. 81

---

15. Which one *best* describes *variation* for a numerical data set?

A. How the data values vary with a single number

B. A single number that summarizes all of the values

C. Interquartile range

D. Mean absolute deviation

---

14. The histogram shows the ages of people at the swimming pool.

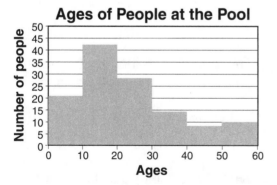

Ages of People at the Pool

About how many people are at the swimming pool?

A. 44

B. 110

C. 123

D. 135

---

16. Which statement about the distribution of the data shown in the box-and-whisker plot is *not* true?

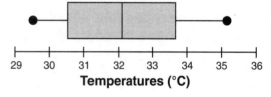

Average Daily Temperatures in June

A. The mean and the median temperatures are about the same.

B. The distribution of temperatures is relatively symmetric.

C. The distribution of temperatures is strongly skewed.

D. There is little variability in the temperatures.

Name: _____ Date: _____

Statistics and Probability:
Multiple-Choice Assessment Prep

Directions: Circle the choice that best answers the question.

---

17. Use the dot plot to answer the question.

Weight of Wrestlers

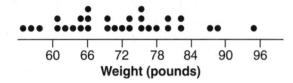

Weight (pounds)

What does each dot represent?

A. The height of each wrestler in feet

B. The height of each wrestler in inches

C. The weight of each wrestler in kilograms

D. The weight of each wrestler in pounds

---

19. Which type of display would best show the distribution of the data given in the frequency table?

Weight (pounds)	Frequency
0 – 2	8
2 – 4	5
4 – 6	7
6 – 8	10

A. Box-and-whisker plot

B. Histogram

C. Dot plot

D. Scatter plot

---

18. Which describes the distribution of a set of data that is collected to answer a statistical question?

A. Center

B. Spread

C. Shape

D. All of the above

---

20. The heights, in inches, of the students on the student council are listed.

49, 52, 52, 54, 57, 62,
66, 67, 67, 67, 68, 71

What is the interquartile range for the heights?

A. 14

B. 22

C. 61

D. 64

Name: _____ Date: _____

Statistics and Probability:
Open-Ended Response Assessment Prep

Directions: Answer the question completely. Show your work and explain your reasoning.

Problem 1: Elton surveys the students in his class about the amount of screen time they have each day. The results are shown in the frequency table.

Screen time (hours)	0 – 1	1 – 2	2 – 3	3 – 4	4 – 5
Frequency	2	7	9	6	3

1. Write an example of a good statistical question that Elton might have used to collect his data.

2. What type of display would be best for this type of data (dot plot, histogram, or box-and-whisker plot)?

3. Create a display of the data.

4. Describe the shape of the distribution and what this means in the given context.

Show your work.

Explain your reasoning.

Name: _____ Date: _____

Statistics and Probability:
Open-Ended Response Assessment Prep

Directions: Answer the question completely. Show your work and explain your reasoning.

Problem 2: The resting heart rates of 23 sixth-grade students are shown.

60, 65, 67, 69, 69, 69, 73, 73, 74, 75, 75, 75, 83, 85, 86, 87, 90, 92, 93, 94, 95, 95, 96

1. Make a dot plot of the data.

2. Describe the distribution of the data.

3. Which measures of center and variability would best describe the distribution?

4. Calculate the measure of center and variability that you chose in Question 3.

Show your work.

Explain your reasoning.

Name: _____ Date: _____

Statistics and Probability:
Open-Ended Response Assessment Prep

Directions: Answer the question completely. Show your work and explain your reasoning.

Problem 3: The lengths, in inches, of 16 babies born on the same day at one hospital are listed.

17, 17, 17, 18, 18, 19, 19, 20, 20, 22, 22, 22, 22, 22, 22, 23

1. Draw a box-and-whisker plot for the data.

2. Describe the center, spread, and overall shape of the distribution.

3. Which measures of center and variability would best describe the distribution?

Show your work.

Explain your reasoning.

Statistics and Probability: Answers and Diagnostics

Multiple-Choice Questions

Problem #	Correct Answer	Diagnostics
1.	C	A. Student chose a single measure of central tendency rather than a description. B. Student chose a single measure of central tendency rather than a description. D. Student chose a description of variation.
2.	A	B. Student chose the number of observations of blue, brown, and green eyes. C. Student chose the number of "other" eye colors. D. Student chose the number of categories of eye color.
3.	C	A. Student chose a plot that can be displayed on a number line. B. Student chose a plot that can be displayed on a number line. D. Student chose a plot that can be displayed on a number line.
4.	D	A. Student did not understand that this question does anticipate and account for variability in the answers. B. Student did not understand how to determine whether or not a question is statistical. C. Student did not understand how to determine whether or not a question is statistical.
5.	C	A. Student did not understand what a box-and-whisker plot is. B. Student did not understand what a histogram is. D. Student did not understand what a dot plot is.
6.	B	A. Student did not understand that mean absolute deviation is a measure of variation. C. Student did not understand that interquartile range is a measure of variation. D. Student did not understand that measures of variation describe how data values vary with a single number.
7.	A	B. Student did not understand how to identify an outlier. C. Student did not understand how an outlier affects the mean. D. Student guessed.
8.	B	A. Student chose Blane's median score. C. Student chose Blane's modal score. D. Student chose the total of Blane's scores.
9.	D	A. Student chose a correct answer, but it is not the *best* answer. B. Student chose a correct answer, but it is not the *best* answer. C. Student chose a correct answer, but it is not the *best* answer.
10.	B	A. Student did not understand which measure of variability to use. C. Student chose measures of center and variability that are best used for skewed distributions and not symmetric distributions. D. Student chose a measure of center that is best used for skewed distributions and a measure of variability that is not paired with that measure of center.
11.	C	A. Student chose the mean/median foot length. B. Student chose the range of foot lengths. D. Student guessed.
12.	A	B. Student did not understand how to analyze the dot plot. C. Student did not understand how to analyze the dot plot. D. Student did not understand what each dot represents.

Statistics and Probability: Answers and Diagnostics

13.	A	B. Student chose the mean number of points.
		C. Student chose the mean of 81 and 77.
		D. Student did not order the data before identifying the middle number.
14.	C	A. Student chose the approximate height of the highest bar.
		B. Student underestimated the total number of people at the pool.
		D. Student overestimated the total number of people at the pool.
15.	A	B. Student chose a description of measures of center.
		C. Student chose a single measure of variation rather than a description.
		D. Student chose a single measure of variation rather than a description.
16.	C	A. Student did not understand the relationship between mean and median for data with a symmetric distribution.
		B. Student did not know how to describe the distribution of data by looking at the box-and-whisker plot.
		D. Student did not know how to determine variability in data by looking at the box-and-whisker plot.
17.	D	A. Student did not read the dot plot correctly.
		B. Student did not read the dot plot correctly.
		C. Student did not read the dot plot correctly.
18.	D	A. Student chose only one of the descriptions for the distribution of a set of data.
		B. Student chose only one of the descriptions for the distribution of a set of data.
		C. Student chose only one of the descriptions for the distribution of a set of data.
19.	B	A. Student chose a display that is not used with categorical data.
		C. Student chose a display that is not used with categorical data.
		D. Student chose a display that is best used with bivariate data.
20.	A	B. Student chose the range of the data.
		C. Student chose the mean of the data set.
		D. Student chose the median of the data set.

Open-Ended Response Questions

Problem #1

1. Sample answer: On average, how many hours of screen time do you have each day: 0 to 1, 1 to 2, 2 to 3, 3 to 4, or 4 to 5?

2. A histogram would be the best type of display for this data because the categories for the possible answers are ranges.

3. **Average Daily Hours of Screen Time**

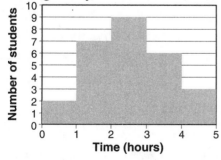

4. The distribution is relatively symmetrical. This means that most of Elton's classmates have between 2 and 3 hours of daily screen time on average. About the same number of students have 1 to 2 hours as those that have 3 to 4 hours, and about the same number of students have 0 to 1 hour as have 4 to 5 hours.

Statistics and Probability: Answers and Diagnostics

Problem #2
1.

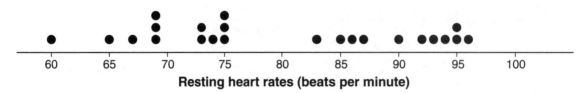

Resting Heart Rates of Sixth Graders

Resting heart rates (beats per minute)

2. The distribution is skewed to the right.
3. Because the distribution is skewed, the median and interquartile range are the best measures of center and variability for this data.
4. 60, 65, 67, 69, 69, (69,) 73, 73, 74, 75, 75, (75,) 83, 85, 86, 87, 90, (92,) 93, 94, 95, 95, 96

         Q1   **Median**   **Q3**

The interquartile range is 92 – 69 or 23.

Problem #3
1.

Lengths of Babies at Birth

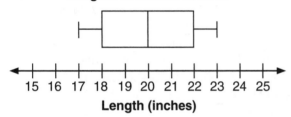

Length (inches)

2. The center is at 20 inches. The spread is relatively even and the shape is symmetrical.
3. Because the shape of the distribution is symmetrical, the mean and mean absolute deviation would best describe the distribution.